333.79
MCL

SALT SPRING ISLAND
MIDDLE SCHOOL

TALKING POINTS
ENERGY CRISIS

Ewan McLeish

Stargazer Books

ABOUT THIS BOOK

What is an ENERGY CRISIS? And how might it affect us? Are we going to run out of energy in the next 40 years, or will there be new sources of energy we can use? These are some of the issues you can read about in this book.

© Aladdin Books Ltd 2008
Designed and Produced by Aladdin Books Ltd

First Published in the United States in 2008 by
Stargazer Books
c/o The Creative Company
123 South Broad Street, P.O. Box 227
Mankato, Minnesota 56002

Designers: Pete Bennett—PBD
Flick, Book Design and Graphics
Editors: Harriet Brown / Katie Dicker
Picture Researcher: Alexa Brown
The author, Dr. Ewan McLeish, is a writer and lecturer in education. He has written over 20 books on science and the environment.
The consultant, Rob Bowden, is an education consultant, author, and photographer specializing in social and environmental issues.

All rights reserved
Printed in the United States

Library of Congress Cataloging-in-Publication Data
McLeish, Ewan, 1950-
　　Energy crisis / by Ewan McLeish.
　　　　p. cm. -- (Talking points)
　Includes index.
　ISBN 978-1-59604-143-1 (alk. paper)
　1. Energy policy. 2. Power resources. 3. Fossil fuels--Environmental aspects. 4. Energy conservation. 5. Renewable energy sources. I. Title
HD9502.2.M37 2007
333.79--dc22
　　　　　　　　　　　　　　　　2007007761

Contents

World Energy Use 6
Learn about how energy and electricity are used in different countries of the world.

Types of Energy 10
Learn all about fossil fuels and the problems that they can cause. Look at some alternative sources of energy.

Renewable Energy 14
Find out about types of renewable energy, such as solar and wind energy.

Fossil Fuels 18
A closer look at oil, gas, and coal. Find out where these fuels are found and how long they might last.

Energy and the Environment 24
Learn about the ways in which using energy can harm our planet.

Saving Energy 30
Look at how we can save energy and make the energy that we have last longer.

Energy Solutions 34
Learn about how we could make and use energy in the future without harming our planet.

Energy in the Future 40
Find out what steps governments are taking to solve the energy problems of the future.

Chronology 44
Find out more about key dates in the history of energy production.

Organizations and Glossary 46

Index 48

INTRODUCTION

In November 2006, 15 million people in Europe were without power. In August 2003, the eastern part of North America also had a serious power failure that trapped people in elevators and trains. These examples highlight what can happen when energy supplies fail.

ENERGY BREAKDOWN

A number of events caused the August 2003 disaster. Lots of air-conditioning systems were switched on because it was a hot day. This meant that there was a great demand for electricity.

The engineers that controlled the energy supply also made some mistakes when using their computer software. Power lines began to shut down and soon the system was out of control.

Other disasters

Since 2003, countries in Europe have also had major power failures. Something seems to be happening to power supplies around the world.

In 2003, 21 power plants shut down in North America leaving 50 million people without power.

Introduction

Four major hurricanes hit the U.S. in 2005 —the worst year on record.

STORM WARNING

Hurricanes are common in parts of the U.S. but in 2005, the country had its worst hurricane season to date. There were 15 hurricanes and four of these were on a major scale. In Hurricane Katrina at least 1,800 people lost their lives. This hurricane caused serious damage—over 80% of the city of New Orleans was flooded for example.

An environmental crisis

We are using a lot of energy very quickly. Scientists think that soon there may not be enough energy to go around. We are in danger of an energy crisis. We also know that burning fossil fuels produces harmful gases that affect the environment. Many people think the weather is changing around the world. This may be linked to the fossil fuels that we are using.

PROBLEM SOLVING

We don't know how long fossil fuels will last or how we will replace them when they run out. We also don't know how damaging our use of fossil fuels will be in the future. But we do know that solving one of these crises could help us to answer the other. This book looks at how we might try to solve these problems.

WORLD ENERGY USE

Recently, a lot of television programs have shown how an energy crisis might affect us. Huge power cuts can have a major impact on the way a country works. We all use energy and we could all be affected.

In developed countries, a lot of energy is used to heat and light homes. Energy is also used to power cars and trucks, and to run our industries.

A tale of two cities

Jan, Norway—Jan is 14 years old. He lives in a big stone and wood house in Oslo, Norway. He is woken at 7:00 a.m. by his radio-alarm, in his bedroom that is centrally-heated. Although it has snowed overnight, snowplows have cleared the roads before the drive to school. Jan turns the light on, showers, and dresses before breakfast. On TV he hears something about a girl called Mala...

Mala, Somalia—Mala is 13 years old. She lives in an iron shack in Mogadishu, Somalia. She is woken by light shining between the metal walls—and because she is hungry. Mala's mother is ill and her father is away looking for work. Mala needs wood for the stove to cook some leftover rice. But wood is expensive. She hopes to pick up some sticks that have fallen from a truck. She dreams of leaving this place—even owning a television...

WORLD ENERGY USE

SPOT THE DIFFERENCE

Look at Jan's story again. How many times does Jan use energy in the morning? You'll be surprised at how much energy he uses.

Now compare this with Mala's story. Energy is an important part of Mala's life, too. But the amount and type of energy she uses is very different.

Electricity used by different countries

Country	kWh
Iceland	26,143
Norway	25,362
Canada	15,661
Sweden	15,194
Finland	14,676
United Arab Emirates	14,126
Kuwait	13,416
Luxembourg	13,365
United States	12,406
New Zealand	8,827
Belgium	7,598
Japan	7,579
Switzerland	7,301
France	6,901

Figures are per person per year in kilowatt hours (kWh).

Energy is important to all countries of the world. However, in developing countries it is more difficult to find energy.

ELECTRICITY USE

Look at the amount of electricity that different countries use in the table above. Iceland and Norway are both cold countries. They make electricity quite cheaply. They use water from rivers and lakes to make "hydroelectricity."

The average American uses about half the electricity used by the average Norwegian. But this is over twice as much as the average Briton or German. In Somalia, Mala uses about one four-hundredth (30 kWh) of the electricity used by the average American. This would power a light bulb for an hour a day.

WORLD ENERGY USE

RICH AND POOR

There are many ways to use energy. Electricity is just one of them. Poor countries often cannot afford to use power plants. Instead, they rely on other forms of energy such as wood and animal dung.

The chart below shows how much energy countries can afford to use. Look at Africa's line in comparison to the other countries.

Energy consumption by economic region (1980-2005)

Legend:
- OECD (e.g North America, Japan, W. Europe)
- E. Europe & former Soviet Union
- Developing Asia
- Middle East & North Africa
- Central & Latin America
- Africa

British thermal units (Btu), 0-225, years 1980-2005

THE POWER OF OIL

In 1990, Iraq invaded a tiny country nearby, called Kuwait. Iraq already produced a lot of oil. However, the country's leader Saddam Hussein wanted more. Kuwait has lots of oil and this would have given Saddam great power.

A number of countries, including the U.S. and Britain, went to war to stop Saddam from taking Kuwait. It was a short, but bloody war and many soldiers were killed or injured. Kuwait was saved, but its oil wells were badly damaged.

In the Gulf War (1990-1991) over 600 oil wells were set on fire.

World Energy Use

An ongoing war

Oil fires burned for many months, sending black, poisonous smoke into the air. The area became polluted. Oil also leaked into the sea.

Oil has now become something to fight a war over. Today, Saddam's reign in Iraq is over, but the fighting still goes on. Countries are very protective about their oil supplies.

The effect of oil

Oil is one energy resource that gives countries power over other countries. Many nations do not have enough energy resources themselves. They have to buy their energy from other countries.

To do so, they also have to remain friendly with these countries. The need for oil is affecting relationships between countries more than ever before.

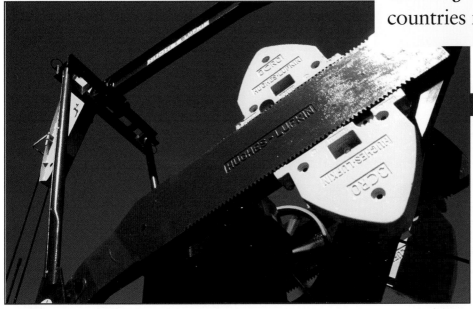

This is an oil pump from Argentina. Countries that do not have very much oil tend to support oil-rich countries, like Argentina, in times of trouble.

Energy in the future

We need to remember that even in rich countries, some people find it difficult to access energy. The unemployed and the elderly, for example, may find it hard to pay their heating bills. Being able to use a lot of energy is also not necessarily the secret to having a good "quality of life." In the future, we may all have to get used to the idea that energy is not something we can take for granted.

Types of Energy

In 1986, a safety test at a nuclear power plant in Ukraine went very wrong. One of the reactors overheated, causing a big explosion. This threw radioactive material high into the air. Many nations nearby were also affected.

The explosion at Chernobyl, Ukraine, polluted large parts of the former Soviet Union and Eastern and Northern Europe.

These solar panels use the sun as a source of renewable energy.

ENERGY USE

Sources of "renewable" energy can be used again and again. The wind, the waves, and the sun are all renewable sources of energy.

Fossil fuels, on the other hand, are non-renewable sources of energy. We're using them up very quickly. This chapter looks at different types of energy sources. It also looks at the problems that they cause.

FOSSIL FUELS

We have used fossil fuels for thousands of years. Fossil fuels are relatively cheap, and easy to store and transport. We also get a lot of useful energy from them.

When fossil fuels are burned, they release harmful gases into the air.

USING FUELS

Most of our energy comes from fossil fuels. Fossil fuels power our cars and we use them to make electricity. Fossil fuels are the most common type of fuel in both rich and poor countries (although rich countries use six times more fossil fuels).

This chart (right) shows that fossil fuels provide over 80% of energy used in rich countries. However, the chart only shows bought energy. Many people also use wood for fuel in poorer countries, for example.

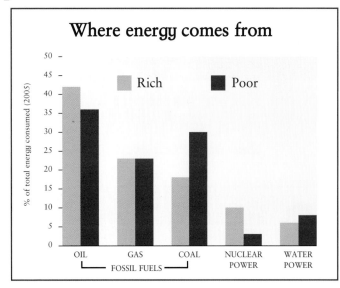

Fossil fuel problems

When fossil fuels are burned, they release gases into the air. Scientists think that one of these gases—carbon dioxide—is causing temperatures to rise. Taking fossil fuels from the ground can also be damaging and dangerous. Many people have been killed in mining accidents. Oil spills at sea have also killed marine life and damaged coastlines. As fossil fuels run out (see pages 18-23), it will become tempting to look for oil in areas, such as Alaska, that are environmentally fragile.

Types of Energy

NUCLEAR POWER

Nuclear power is made by splitting uranium atoms. The energy holding each atom together is released as heat. The heat can be used to turn water into steam. The steam then turns machines that make electricity.

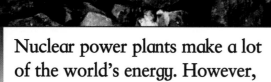

Nuclear power plants make a lot of the world's energy. However, they are expensive to build.

? Highs and lows

When nuclear power became available in the 1950s, most people thought it would make electricity very cheap. Nuclear power was also popular because it didn't give off many polluting gases.

However, nuclear power turned out to be far more expensive than coal or oil. People were also worried about it being dangerous.

A growing problem

There are over 443 nuclear power plants in the world, in over 30 countries. Many old power plants will have to close down. This needs to be done very safely.

Safety issues

Nuclear accidents (see page 10) don't happen very often. However, nuclear power produces a lot of radioactive waste. This remains dangerous for thousands of years. Many old power plants are unsafe to use. Closing them down is expensive and dangerous.

In some rich countries, nuclear power provides around 17% of electricity. The worry of fossil fuels running out (and their effect on the environment) has made many countries look at nuclear power again. However, other countries are planning to close down their nuclear power plants. This is because of the cost and possible dangers.

THE THREE GORGES DAM

A new dam is being built on the Yangtze River in China. The Three Gorges Dam should supply 10% of China's energy. This is a type of "hydroelectric" power plant. The dam causes a large lake to form. The water is then used to turn big turbine blades to make electricity.

Big or what!

The Three Gorges Dam is due to be finished in 2008. It is nearly 1.5 miles (2.5 km) long. The dam will create a lake 390 mi (640 km) long. Many people have had to move their homes for the project.

Hydroelectricity

The energy from falling water helps to make a sixth of the world's electricity. Hydroelectric power plants are cheap to use. They also produce very few polluting gases. The power plants can help to control flooding. Using hydroelectric power means that less wood is needed for fuel.

Disadvantages

Building a big dam is very expensive. A lot of energy and building materials are used in the process.

Dams also flood the land in the area. Often this land is used by farmers or is the home of rare wildlife. Dams start to work less well over time because the lakes fill up with mud and sand. Flooded forests can also produce harmful gases such as methane and carbon dioxide.

Many local people are unable to afford the energy produced by hydroelectric dams. They still have to use wood from local forests.

RENEWABLE ENERGY

In the hills above San Francisco, a group of 300-ft (100-m) giants tower over the ground. They have bladelike arms that move in the wind. These blades are used to produce electricity. Other sources of renewable energy include the sun, the sea, and heat from the earth.

This wind farm near San Francisco has 6,000 big wind turbines that make electricity.

World wind power

Europe makes the most wind power in the world. It produces nearly three-quarters of the world total. In 2006, wind power made less than 1% of energy used around the world. By 2020, this figure is expected to rise to 12%.

GOOD AND BAD

Wind power is one of the fastest growing renewable energies. Many people don't like wind turbines because they think they look ugly and make a lot of noise.

People are also worried about their impact on birds.

Putting wind turbines out at sea is one solution to these problems. The first in the U.S. could begin to generate electricity in 2009.

THE POWER OF THE SUN

We can use energy from the sun in two different ways. We can use the sun's heat energy for solar heating or we can turn its radiation into electricity. The radiation is turned into electricity using a special device called a photovoltaic (PV) cell.

Using PV cells

Groups of PV cells are connected to make more power. Lots of buildings now have PV cells (above). Recently, some very large PV power plants have also been built in the U.S. and Europe.

THE POWER OF THE SEA

There is an endless supply of wave power. Wave power plants have been built close to coastlines around the world. Tidal energy can also be used to generate electricity. Turbines are put on large "barrages" built across estuaries. One of these barrages can be found on the Rance River in France.

IN THE FUTURE

There are also plans for huge barrages built to produce a lot of power for homes. One of these is planned to cross the UK's Severn Estuary. It would be 10 miles (16 km) across and could produce up to 5% of the UK's electricity!

Large barrages, however, have disadvantages. Stopping the tide can seriously affect the feeding patterns of wildlife nearby.

A wave power plant
Waves enter (1); rise and fall in a column (2); air is pushed up and down (3), moving a turbine (4) to work a generator that produces electricity.

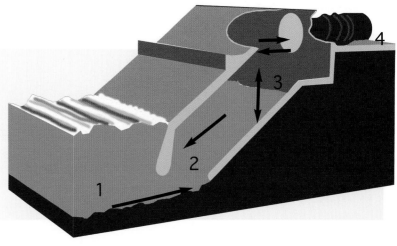

BIG AND SMALL DAMS

Not all hydroelectric power plants are as big as the Three Gorges Dam (see page 13). Small dams already produce up to a third of China's electricity. The electricity they produce is used nearby. The dams also cause little damage. Small hydroelectric dams work well in countries with lots of mountains and rivers.

HEAT FROM THE GROUND

In some parts of the world, there is enough heat underground to heat water and steam up to 480°F (250°C). This geothermal (heat) energy can be used to make electricity.

Geothermal leaders

The U.S. generates the most geothermal power, followed closely by the Philippines. The remaining top five countries are Mexico, Italy, and Indonesia. In 2005, the U.S. generated over 2,300 Megawatts of geothermal power.

The heat from geothermal reservoirs is used to produce electricity at geothermal power plants.

Geothermal energy

New geothermal technology is being developed all the time. Cold water is pumped into hot rocks, for example, where it is heated so it can be used.

Heat pumps are also used to extract heat energy from depths of 320-490 ft (100-150 m). These pumps don't provide very much energy overall, but they are useful for many rural homes in the U.S. and Europe.

Biofuels, such as sugar cane, are said to be "carbon neutral." The carbon they release when they are burned is usually equal to the carbon they use when they are growing.

BIOFUELS

Biofuels are materials such as wood, straw, or animal wastes that were once living things (biomass). This material can be burned to produce energy.

In many poor countries, wood is a common source of energy. Wood is taken from forests and farmland. Animal dung can also be used as a fuel. However, farmers want to use dung as a fertilizer.

Some biofuels are produced in large amounts. Crops such as sugar cane are grown in countries like Brazil. Waste such as straw or household trash can also be collected.

A QUESTION OF CONTROL

Renewable sources of energy can be used on a small scale for local people. Once built, the technology is free to use and people will have plenty of energy available. We must remember, however, that nonrenewable fuels are not always "bad" and renewable fuels are not always "good." All energy sources have advantages and disadvantages.

Wood is a useful biofuel. However, trees take a long time to grow.

FOSSIL FUELS

In recent years, China has used a lot of oil and coal. This is because the country is developing. Some people think that if countries develop in this way, we are going to run out of fossil fuels very quickly.

FUEL SOURCES

We know that some countries have more energy than others. Now we need to look at the countries and areas that control the flow of energy around the world. The chart on the left shows the amount of fuels that are produced. You can see that fossil fuels are produced the most. Nuclear and hydro-electric power come equal fourth. Various types of renewable energy are last.

Coal is the second most widely used source of energy. When coal is burned it releases carbon dioxide into the air.

World energy production

Now look at the table below. This shows which areas of the world produce energy. The Americas are well in the lead, followed by Asia and Oceania, and then Eastern Europe and the former Soviet Union. Although Western Europe uses a lot of energy, it produces relatively little. This means it needs energy from other countries.

The U.S. produces the most energy, followed by Russia and China. This order of countries is likely to change, however, as fossil fuels begin to run out.

Middle Eastern states produce more oil than any other part of the world.

OPEC

Many of the world's leading oil nations belong to a powerful group called the Oil and Petroleum Exporting Countries (OPEC). They control the supply of about half the world's oil.

Oil prices are always changing. In 1973, oil became very expensive (see page 20). Since 2004, the price has continued to rise further. Oil prices have risen because oil is in danger of running out. Although oil production has increased in recent years, prices are still high. This is because many oil-producing countries are unstable.

Fossil Fuels

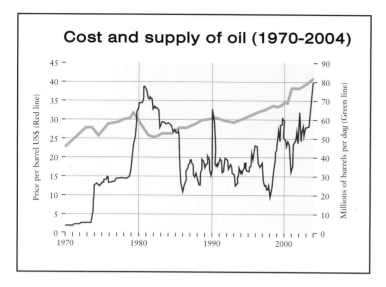

ENERGY USES

Energy is used in many different parts of the world. We use energy for a number of reasons. In the U.S., for example, 38% of the country's energy is used in industry, 35% is used for homes and offices, and 27% is used to move people and goods from place to place.

Industry uses a lot of energy, particularly in developed countries.

EUROPEAN TRANSPORT

In Europe, cars and trucks use the most energy. Air travel is the second greatest transport energy user. Although airplanes use a lot of energy to fly, there are less planes than cars. In Denmark, for example, the energy used by vehicles on the road increased by 50% between 1970 and 2000.

In Western Europe, 85% of transport energy is used by people driving in cars. Less energy is used in buses and trains. This is partly because buses and trains need less energy to work. In Europe, most industrial goods are transported by road. Trains use less energy, but trucks make it easier to get to small places.

INDUSTRY

About 70% of energy used in European industry goes toward manufacturing. This is followed by agriculture (farming) and construction (building). However, things are very different in other parts of the world. Poorer countries rely on farming. Farming needs animal or people power (rather than machinery). But as these countries develop, their energy needs change.

ENERGY AT HOME

Most European homes use about 60 kilowatt hours of energy a day. Most of this is used in heating. This energy mainly comes from fossil fuels (or electricity made using fossil fuels). In poor parts of the world, wood or a liquid called kerosene are often used. However, as kerosene doesn't burn very well, it wastes energy and releases harmful gases.

Space heating 58%
Cooking 5%
Lights and appliances 13%
Water heating 24%

Energy use in a typical European home

FUEL SUPPLIES

Scientists hope that we'll be able to find alternative sources of energy before we run out of fossil fuels completely. This depends on how much energy reserves we have left.

The chart on page 22 shows that the Middle East has a lot of oil and gas. The Americas also have some oil reserves. There are still large gas reserves in the former Soviet Union and Eastern Europe. Coal reserves, on the other hand, are found in more places. Coal is mainly found in the U.S., Russia, China, India, and Australia.

Most of the energy we use at home or at school comes in the form of electricity. This is generated in power plants using oil, gas, coal, or nuclear power.

RUNNING OUT OF TIME

We currently have about 1,147 billion (1.1 trillion) barrels of oil left. We use 75 million barrels of oil a day. If we go on using oil at this rate, we only have about 40 years' worth of oil left. Most scientists think that gas will last for 70 yrs and coal for 190 yrs.

Finding fuels

It is not easy to estimate the amounts of fuel in the ground. We normally talk about "known" sources. There may be other sources of fuel that we don't know about. As fossil fuels begin to run out they will become much more expensive.

If we go on using oil at today's rate, it is likely to run out in your lifetime.

Slowing down

We are getting better at finding fuels. And if we start to use less energy, fossil fuels will last longer! However, as current fuel reserves run out, other reserves will be more difficult to find. In about 30 years, fossil fuel production is likely to slow down.

Sources of crude oil and natural gas

Making changes now

It looks as though oil and gas will have run out (or become scarce) in about 50 years. Since most oil is found in the Middle East, countries will rely more heavily on this part of the world for their energy. This situation could be improved in the future if we start to save oil and gas now.

In the future, we need to find alternatives to fossil fuels.

Relying less on fuels

One way to do this would be to only use fossil fuels where really necessary, and to start using alternative types of fuels.

Coal will continue to be an important fuel for many years. This is particularly true in developing countries such as India and China. But relying on coal could be damaging to the environment.

Although coal is now treated, it still produces more carbon dioxide than oil or gas.

ENERGY AND THE ENVIRONMENT

In 2006, a 200-ft (60-meter) wall of ice fell from a glacier in Argentina. In 2002, the hanging Kolka Glacier in Russia (right) also collapsed. Glaciers are melting around the world.

Although scientists thought that large glaciers might collapse as the climate warmed, no one had predicted the 2002 Kolka Glacier tragedy. An avalanche buried villages and over 100 people were killed.

CLIMATE CHANGE

Temperatures around the world have risen over half a degree in the past 150 years. Scientists also think that this increase is speeding up. Many people believe the effect on world climate could be huge.

However, the cause of climate change is still debated. The Earth's climate changes over periods of time. But many people think that our way of life is affecting this temperature change.

Mount Dzhimarai-Khokh before (left) and after (right) the collapse of the Kolka Glacier. These images were captured by astronauts in the International Space Station.

GLOBAL WARMING

About 24% of the heat and light energy that reaches the earth from the sun travels through the earth's atmosphere. This energy heats the earth's surface and is then reflected back into space.

However, an increase in "greenhouses gases" in the atmosphere, such as carbon dioxide, has now changed this pattern. Greenhouse gases absorb some of the heat energy reflected from the earth's surface. They then reflect some of this energy back to the earth. This means that the planet gets warmer.

The effect of carbon dioxide

The graphs below show the rise in temperature and carbon dioxide levels over the past 100 years. The lines look very similar. Most scientists think that this is because the two things are related.

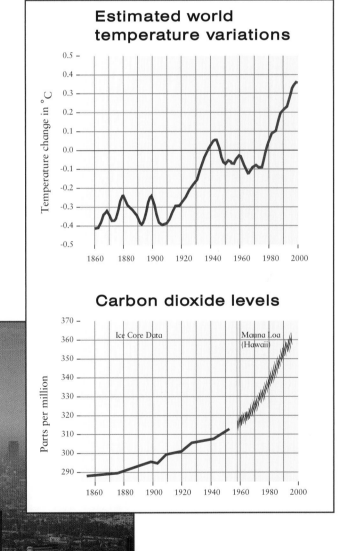

Pollution from cars causes cities like Los Angeles to be blanketed in smog.

ENERGY AND THE ENVIRONMENT

Record highs

Every year we release about seven billion tons of carbon into the air. This carbon was buried millions of years ago. It will now remain in the atmosphere for about 100 years. Carbon levels have been increasing as countries develop.

RISING SEA LEVELS

Sea levels have risen by about 0.8 in (2 cm) over the past 100 years. Scientists estimate that at the present rate of warming, sea levels could rise another 23 in (60 cm) by the end of the century. This would make it impossible to live in many low-lying areas of the world. Countries like Bangladesh would be at risk of flooding and many Pacific islands would disappear.

Rising sea levels are partly caused by ice melting at the North and South poles. Antarctic glaciers, for example, are thinning twice as fast as they were in the 1990s.

If sea levels rise by 23 in (60 cm), cities such as Tokyo and New York could become flooded.

Other gases

Methane, nitrogen dioxide, and ozone are also greenhouse gases. Methane is produced by bacteria. The bacteria like to live in places where there is little oxygen (such as in landfill sites). Methane levels are increasing as temperatures rise. Greenhouse gases like nitrogen dioxide and ozone are produced by cars and trucks.

Ice in the Arctic contains frozen methane. If temperatures continue to rise, this gas may be released.

ENERGY AND THE ENVIRONMENT

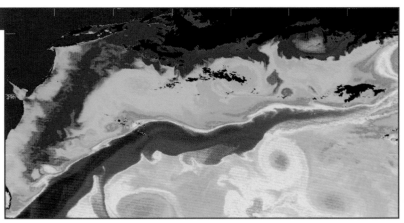

The red strip in this image shows the Gulf Stream. This is a large area of warm water in the Atlantic Ocean.

THE GULF STREAM

Much of Western Europe is kept warm by the Gulf Stream. This is a huge ocean current that moves warm water around the oceans. The Gulf Stream is kept going by a deeper current, called the Conveyor, that moves in the opposite direction. As more icecaps melt, the Conveyor could slow down or stop altogether. This would affect the Gulf Stream. Parts of the Northern Hemisphere would then cool very quickly.

WINNERS AND LOSERS

Areas farthest from the equator are likely to see the greatest temperature rises. Changing weather patterns and water shortages will influence farming. Crops will fail in some parts of the world and begin to grow faster in other areas. Overall, the effects of climate change could be disastrous.

Meltdown!

The Greenland Ice Cap is about 7,000 ft (2,135 m) high and contains around 0.6 million cu mi (2.5 million cu km) of ice. Over the past 30 years the ice cap has reduced, particularly in the warmer southwest. A total melting of it would raise sea levels a massive 24 ft (7.4 m)!

Agricultural patterns around the world could change if temperatures rise.

Deadly chain

For centuries, groups of seabirds have lived on the islands of Northern Scotland. However, in 2004, the cliffs fell strangely silent as thousands of birds failed to raise any young. The temperature of the North Sea has risen by 4°F (2°C) in the last 20 years. This has affected the types of food on which the birds feed.

CHANGING LIVES

Insect-eating birds, like titmice, normally produce their young when there are plenty of caterpillars on which to feed. Warmer springs mean that caterpillars hatch much earlier. The birds may change their breeding cycle to follow the caterpillars. However, some scientists think they will not be able to change quickly enough.

Moving away

As temperatures change, wildlife may have to move north and south. Some plants and animals may change, but scientists think that at least 9% of species face extinction.

GAS PRODUCERS

The chart on page 29 shows which countries produce most greenhouse gases. Asia and Oceania are first, followed by North America and Western Europe. The U.S., China, and Russia are the countries that produce the most carbon dioxide.

Many creatures are facing extinction around the world. Only some species will be able to change to cope with the effects of global warming.

Carbon dioxide producers

The energy needs of countries like China, Indonesia, Thailand, and Korea are rising very fast. Carbon dioxide production has increased in these countries in the past 20 years.

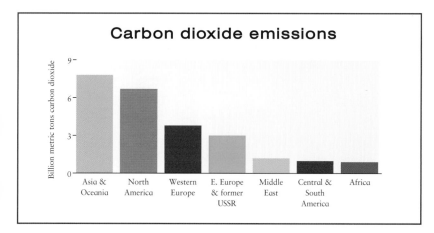

ACID RAIN

When fossil fuels burn, the gases they release can cause rain water to become acidic. Acid rain damages forests, lakes, and buildings. In western countries, reducing the amount of coal used has helped to reduce levels of acid rain. However, as other parts of the world develop, acid rain levels are likely to increase again.

KEEP THE TREES

Cutting down trees has a major impact on carbon dioxide levels. Trees absorb carbon dioxide and store the carbon for many years. Burning trees releases the carbon back into the atmosphere.

FUTURE COSTS

Our planet could be seriously damaged by our use of fossil fuels. But there are other costs, too. If the climate changes, food supplies could be affected and farming countries may become poorer. The poor are most likely to suffer because it is more difficult for them to change their situation.

Acid rain damages forests, soils, and wildlife.

Saving Energy

In most power plants, steam is used to turn a turbine. The turbine then works a generator which produces electricity. However, a lot of heat energy is wasted in this process.

These men are making a heat exchanger. These are useful for industries because they reuse waste heat.

WASTED HEAT

The steam in a power plant is heated to about 1,200°F (650°C). At this temperature it works very well. However, using a liquid that boils at a lower temperature than water requires less energy to heat the liquid.

Propane, for example, has a low boiling point (around 120°F or 50°C). Using it could make power plants up to 60% more efficient. Although waste heat would still be produced, it could also be reused.

Doing more with less

Unless we develop better ways of using energy we are likely to have an energy crisis. A lot of people are trying to save energy in their daily lives. You may have "energy-efficient" light bulbs in your home or at school. Other items at home, such as washing machines, have to meet strict energy guidelines. It is important that we start saving energy now.

Saving Energy

ENERGY EFFICIENCY

Whenever energy is used, it is turned from one form (such as heat) to another (such as electricity). Every time energy is changed from one form to another, some energy is wasted. The more times the energy is changed, the greater the amount of energy that is wasted.

Saving energy in industry

In rich countries, industry uses more than one-third of a country's energy. Most of this energy comes from gas, oil, and nuclear power.

Some industries use their own waste as a source of fuel. Keeping equipment in good working order and using insulation can also help to save energy.

A lot of energy changes take place when appliances are made (or used!)

The steel industry uses a large amount of energy.

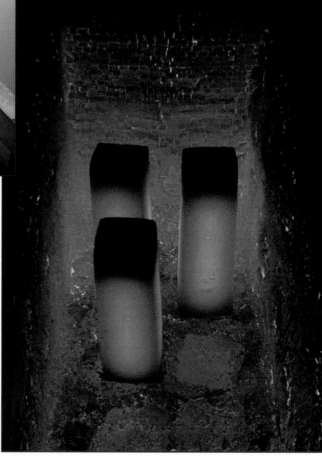

Ways to heat water

If you heat water using electricity, about five energy changes take place. However, heating water using solar heating for example, uses just one energy change. We find it useful to use electricity, but it is wasteful to use it if we only want to heat water to low temperatures.

ENERGY AND TRANSPORT

Cars, trucks, and planes use the most "transport" energy. A car uses nearly four times as much energy as a bus or a train to carry one person 0.6 miles (1 km).

Gasoline or diesel-driven cars use less fuel than they used to. They are less polluting. New materials make these vehicles lighter. "Catalytic converters" also remove poisonous gases from their exhausts.

Transport pollution

In some parts of the world, many older cars do not have catalytic converters. These cars produce polluting gases. Our use of cars is also increasing.

Battery-driven vehicles are clean and efficient, but they need to be charged with electricity. New types of Hybrid cars are now being developed. These use both gasoline and electricity. There are also plans to improve public transportation to make journeys more efficient.

Airplanes use up to 11 times more energy than a bus or a train to carry a person 0.6 miles (1 km).

Car manufacturers are now starting to produce hybrid engines as well as battery-operated vehicles (see page 35).

This building in Massachusetts is energy-efficient. It has 3,000 windows and low-energy systems.

HOME COMFORTS

Most new homes and offices in the U.S. or Europe are built using strict energy guidelines. Many older buildings, however, lose a lot of heat energy.

In 2004, a campaign was launched during Energy Efficiency Week in the UK (see below). It highlighted five things people could do to save energy.

Five energy savers

- Turn down the thermostat by 3°F (1°C).
- Use low-energy light bulbs. (If everyone changed one light bulb in their house, millions of dollars would be saved).
- Turn televisions off instead of on standby.
- Turn off lights in empty rooms.
- Boil just enough water for your needs.

THE PRICE OF OIL

The price of oil is controlled by some of the oil-rich countries. However, there are extra costs to using oil. Transporting, refining, and delivering oil all cost money. Some countries, such as the UK, also add a lot of tax. They do this to encourage people to use public transportation instead of cars. A gallon (3.8 liters) of fuel currently costs three times more in Europe than in the U.S.

In 2000 and 2005, roads in the UK were blocked by truck drivers and farmers. They were angry that high fuel prices were causing them to lose business. However, the government still refused to lower the tax on gasoline and diesel.

ENERGY SOLUTIONS

In southwest Australia there is a tower that is 0.6 miles (1 km) tall. This is three times taller than the Eiffel Tower in France. It stands at the center of a sea of glass and plastic 4 miles (7 km) across.

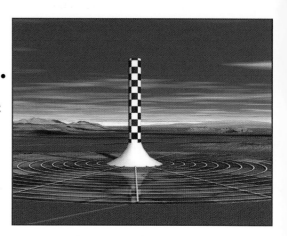

At the moment, this tower doesn't actually exist. But one day, designers and engineers hope that it will reach up into the sky and suck up renewable energy from the sun.

THE POWER TOWER

If the Power Tower is built, it will be the world's biggest solar power plant. Air heated by the sun under its giant plastic skirt will rise up the tower, turning 32 turbines. This will generate enough energy to supply 70,000 people every year.

Designers hope that a Power Tower could eventually provide electricity to poor nations with plenty of sunshine. Others think the tower would be ugly and expensive to build and use.

Power Tower 3,280 ft (1,000 m)

CN Tower 1,814 ft (553 m)

Eiffel Tower 1,062 ft (324 m)

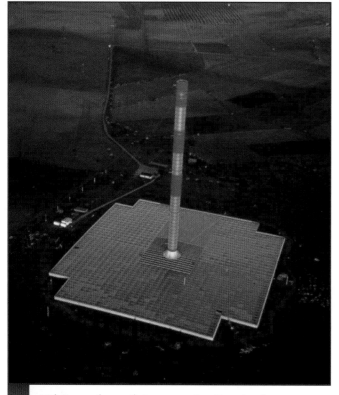

This solar chimney in Spain has a greenhouse, a chimney, and a turbine to generate electricity.

JOINING ATOMS

Splitting atoms of uranium (see page 12) is called fission. This can be used to work turbines but it produces dangerous radioactive waste. Atoms can also be joined in a process called fusion. This is similar to the way our sun produces energy. Scientists have joined atoms in a laboratory and there are plans to build a reactor in the future.

HOPE FOR HYDROGEN

Hydrogen gas can be burned as a fuel. However, it is better to change the energy it contains into electricity. Fuel cells take electrons from hydrogen atoms and use them to create an electric current. The electrons are then mixed with hydrogen and oxygen to produce water. Water is the only waste product.

This swimming pool is powered using a hydrogen fuel cell.

Hybrid vehicles

Cars powered by fuel cells would need "hydrogen stations" instead of gas stations. At the moment, hybrid cars are a better alternative. These have a gasoline engine which recharges an electric engine. This halves the amount of fuel used.

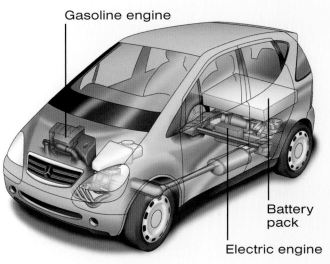

Further sources of hydrogen

Hydrogen can be extracted from methane (a gas produced from waste material). However, carbon is produced as a waste product. Another source is water which can be split into hydrogen and oxygen using electricity. But this process is very expensive. Using renewable energy, such as wind power, would be cheaper, and less polluting.

ENERGY FROM THE TIDES

In September 2003, a large underwater power plant was linked to an electricity supply for the first time. This power plant is off the coast of Norway. It looks like a large windmill and captures tidal currents which turn turbines.

Now, new tidal power plants are being developed. Soon there may be 20 tidal mills off the coast of Norway. The European Commission has also found 106 possible places for tidal power plants around Europe's coastlines. The power of waves is also useful. However, the energy in waves is more difficult to capture (see page 15). Wave power may take longer to develop.

Marine turbines are driven by ocean currents. They do not harm fish because the blades turn slowly —around 20 times per minute.

Better than barrages!

Using tidal currents solves many of the problems caused by tidal barrages (page 15). A circular dam can also be used to trap water at high tide. It could then be allowed to escape over turbines as the tide falls. These designs are less damaging than barrages.

A PLACE IN THE SUN

PV cells (see page 15) can convert 36% of the sun's radiation into electricity.

PV cells are not very efficient at the moment. However, it is possible that solar power could supply over a quarter of the world's electricity by 2040.

ENERGY SOLUTIONS

Many countries, including the U.S., Germany, and Japan, have solar heating and PV cells in homes and offices.

Solar-powered buildings

Buildings of the future could be clothed in a flexible material that absorbs radiation from the sun. The flexible panels would have a large surface to absorb lots of sunlight.

WIND FOR THE FUTURE

For many people, wind turbines are still the best option. Wind turbines have become cheaper, quieter, and more efficient. Roughly 17% of the world's wind power is produced in the U.S.

City life

In the future, buildings may have in-built turbines (see the illustration below). They would be built like curved towers which would draw wind toward the turbines. Although wind speeds are reduced in cities, the turbines could produce at least 20% of the building's energy needs.

Energy Solutions

Out at sea

Using wind turbines at sea can prevent some of the problems of wind farms. "Offshore" winds tend to be stronger than those on land. In 2002, the largest offshore wind farm was completed at Horns Rev in Danish waters. An even bigger project is now due to be built in the Irish Sea, 12 mi (20 km) south of Dublin.

The 80 turbines at Horns Rev in Danish waters generate enough electricity for 150,000 Danish homes. Scientists estimate that wind power could supply more than 20% of the world's energy by 2040.

WATER POWER

Large-scale hydroelectric power plants (see page 13) could provide the whole world's electricity needs. However, in most countries, the rate of development of water power plants has fallen behind other energy technologies. Rich countries have run out of places to put them. And poor countries have found that they have spent a lot of money on hydroelectricity, but the benefits have not been very big.

Small or large

Small dams solve some of the problems of large hydroelectric power plants. They can be built in remote areas to help people living nearby. At the moment, China is the only country to really benefit from a large dam (see page 13).

Large dams have become a less popular way to produce energy.

CHANGING ATTITUDES

In some ways, there are no "new" sources of energy. We are just finding better and cheaper ways to use energy sources that have always been there. We don't know which energy sources will best replace fossil fuels in the future. In fact, it is likely we will need them all.

The solution to our energy problems is also a question of attitude. It is easy for most of us to take energy for granted. This is especially true if we don't have to pay any fuel bills! We need to start thinking about energy in a different way and to start changing attitudes today.

The attitude of young people could be one of the most important ways to prevent an energy crisis.

Energy in the Future

In 2006, an increase in the price of oil caused countries to think again about using nuclear power. France is currently the only major country to produce more than half its electricity using nuclear power.

Many countries are now trying to reduce their carbon dioxide emissions.

THE KYOTO AGREEMENT

In 1997, a number of countries made an agreement at Kyoto, Japan. They promised to release 5% less carbon dioxide on average (compared to 1990 levels), by the year 2012. Many of the world's poorer countries did not sign the agreement. This was because they needed to develop their industries further.

BACKING KYOTO

Kyoto was the first time that countries had chosen to come together to agree to cut down on fossil fuels. The world was beginning to accept that global warming was a threat.

However, not everyone agreed. The U.S. and Russia felt that there was not enough evidence to make them change their policies. Without the support of at least one of them, the Kyoto agreement could not happen.

If we use less energy today, people will benefit in the future.

Only the beginning

Then in October 2004, Russia changed its mind. The country had reduced its levels of carbon dioxide and it wanted to be rewarded. Kyoto could finally go ahead, but there was still a long way to go. Today, carbon dioxide levels are still high and the U.S. has yet to sign up to the agreement.

Beyond Kyoto

Some people think that nuclear power is the answer to rising carbon dioxide levels. Others think that it is a dangerous and outdated technology which should be avoided.

As the debate goes on, carbon dioxide levels keep rising. They currently stand at around 380 ppm (parts per million). If levels rise to 400 ppm, temperatures will increase by another 2.8°F (1.5°C). At current rates this could be within 10 years!

LIVING WITHOUT OIL

Our world needs fossil fuels, especially oil. We will probably still use oil for the next 25 years as our energy use rises (see chart below). By this time, oil may be running out and global temperatures may be rising fast. However, if we develop renewable energy sources they could be used to fill the "energy gap." And the benefits would be huge.

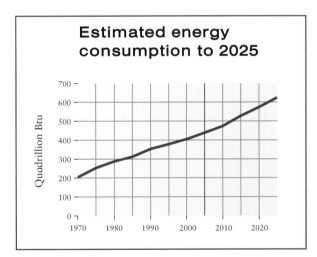

Carbon dioxide pollution has now reached a record high.

IS IT POSSIBLE?

Studies show that simply heating, cooling, and lighting buildings more efficiently could greatly reduce our use of energy. Using hybrid engines (see page 35) would also double the efficiency of our cars. And wind and solar energy would be a great help.

Scientists think that if we do these things, we could stop the increase of greenhouse gases by 2050. Developing energy technologies would also create millions of jobs worldwide.

If we made our buildings more energy efficient we could reduce the amount of fuels that we use.

PLANS FOR THE FUTURE

The Kyoto agreement was a good start for the future. Another recent idea hopes to help even further.

The plan is for richer countries to reduce their levels of greenhouse gases to match those of developing countries who are still expanding. Greenhouse gas emissions worldwide would need to be reduced by at least 60% by 2050 to keep carbon dioxide levels below 450 ppm.

The use of renewable energy sources, such as solar power, is one way to reduce carbon dioxide emissions.

ENERGY IN THE FUTURE

Putting a price on carbon

On January 1, 2005, the European Union passed the biggest environmental law in history. It put a price on carbon. The scheme allows "carbon allowance trading," so companies can buy or sell their allowances. The measures are meant to encourage companies to introduce better ways to save energy.

WHAT THE FUTURE HOLDS

World problems are often linked. If you solve one problem you sometimes solve another. The greatest problem of the 21st century is to find new sources of energy and to prevent the effects of pollution, such as global warming.

In the future, renewable energy sources could help us. They could replace fossil fuels and are kinder to the environment. Governments are now taking the problem seriously. However, there are already signs that targets are slipping. We cannot afford for this to happen. We are all responsible for our planet's future. We should never make the mistake of thinking it is someone else's problem.

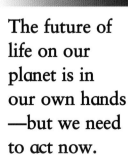

The future of life on our planet is in our own hands —but we need to act now.

Chronology

1.6 million BC—Early man (Homo erectus) learned how to make fire.

1500 BC—Hot springs were used for bathing, cooking, and heating by the Romans, Japanese, and others.

200 BC—The Chinese first began to mine coal.

1100—Windmills were first introduced to Europe.

1740—Commercial coal mining began.

1821—The first natural gas well was drilled in New York.

1834—Thomas Davenport invented the electric streetcar (the tram).

1839—The first fuel cell was made by Sir William Robert Grove.

1859—The first oil-production well was drilled in Pennsylvania.

1860—Etienne Lenoir invented the first gasoline engine.

1880—Coal-fired steam generators were used to generate electricity for the first time.

1882—The first hydroelectric dam was built in the Tennessee River Valley.

1885—Karl Benz invented the gasoline-driven car.

1887—An American, Charles Brush, built the first automatically operating wind turbine to produce electricity.

1891—The first natural gas pipeline was built from Indiana to Chicago. It was 120 miles (193 km) long.

1893—A Frenchman, Abel Pifre, designed a solar engine and used it to run a printing press.

1903—Wilbur and Orville Wright built the first powered aircraft.

Chronology

1904—The first geothermal power plant (producing electricity) was built in Larderello, Italy.

1938—Otto Hahn and Fritz Strassman showed how nuclear fission (splitting the atom) works.

1944—The first nuclear reactor began working in Richland, Washington.

1954—D.M. Chaplin and others invented solar (PV) cells.

1973—Oil prices increased due to fears that fossil fuels were running out (and were also harming the planet).

1979—A nuclear power plant at Three Mile Island, Pennsylvania, came within 30 minutes of a major explosion.

1986—The Chernobyl Nuclear Reactor, in Ukraine, suffered a meltdown, releasing large amounts of radioactive material.

1997—The Kyoto Protocol was signed. Over 160 countries agreed to cut their carbon dioxide emissions.

1998—Buses powered by hydrogen (fuel) cells joined the Chicago mass transit system.

2003—A serious power failure struck North America. This started fears of an energy crisis.

2004—Russia finally signed up to the Kyoto Protocol, allowing the agreement to go ahead.

2005—The European Union brought in carbon allowances for member countries to help meet Kyoto targets.

2006—The U.S. and Russia united in support of nuclear energy at a G8 meeting. Gasoline prices reached $3.00 per gallon in the U.S.

2007—EU leaders agreed to reduce carbon emissions by 20% by 2022 (and by 30% if other rich nations such as the U.S. signed up).

Organizations and Glossary

American Solar Energy Society
www.ases.org

The ASES encourages the use of solar energy.

American Wind Energy Association
www.awea.org

The AWEA's aim is to promote the growth of wind power by raising public awareness through communication and education.

Association of Power Producers of Ontario (APPrO)
www.appro.org

The main site for Canadian power producers. Focuses on producing electricity in efficient and environmentally friendly ways.

Canadian Renewable Energy Network (CanREN)
www.canren.gc.ca

Aims to increase the understanding of renewable energy so that new technologies can be developed for the future.

Canadian Wind Energy Association
www.canwea.ca

The CWEA's purpose is to provide clean and reliable energy throughout Canada and to encourage investment in renewable energy sources.

Interstate Renewable Energy Council
www.irecusa.org

Educates people about renewable energy and its future benefits.

U.S. Department of Energy
www.eere.energy.gov

Provides a gateway to websites and online information about energy efficiency and renewable energy.

Other useful websites:

www.cdea.ca
Canadian District Energy Association.

www.centreforenergy.com
Canada's primary source of information on energy-related issues.

www.eia.doe.gov/emeu
U.S. Energy Information Agency.

www.greentreks.org
A U.S. company that works to improve the environment and reduce pollution.

Organizations and Glossary

Barrages—Dam-like structures built across estuaries. They trap the energy of tides and change it into electricity.

Biomass (bioenergy)—Energy we get from living matter, such as wood, straw, or animal waste.

Climate—The normal weather patterns of an area.

Emissions—Gases, such as carbon dioxide and sulfur dioxide, which are produced when fuels are burned.

Fossil fuels—Fuels we get from the fossil remains of tiny plants and animals. Fossil fuels include coal, oil, and gas.

Fuel cell—A device that makes electricity from hydrogen and oxygen. Water is the waste product.

Geothermal energy—Energy we get from the natural heat of rocks.

Global warming—The warming of the earth caused by the buildup of greenhouses gases (see below).

Greenhouse gases—Gases such as carbon dioxide and methane which absorb and reflect heat energy from the earth's surface.

Hydroelectricity—Electricity generated by using the energy of running water.

Insulation—A material that helps to keep buildings warm and reduce energy use.

Nuclear fission—Splitting atoms and using the heat energy released to produce steam to work turbines.

Nuclear fusion—Joining atoms together at high temperatures to generate energy.

Photovoltaic (PV) cell—A device that changes energy from the sun into electricity (also known as a solar cell).

Renewable energy—Energy such as wind and solar power that does not run out when we use it.

Solar heating—Using the warmth of the sun's radiation directly to heat water (or air) in a building.

Turbine—A machine with blades, that is turned by steam, water, wind, or another source of energy. Turbines are attached to generators which produce electricity.

Index

barrages 15, 36, 47
biomass 17, 47

carbon dioxide 11, 17, 18, 23, 25, 26, 28, 29, 40, 41, 42, 43, 45, 47
climate change 5, 11, 24, 27, 41, 47
coal 11, 18, 21, 22, 23, 44

dams 13, 16, 38, 39, 44

electricity 4, 7, 8, 11, 12, 13, 14, 15, 16, 18, 21, 30, 31, 34, 36, 38, 39, 40, 44, 45
emissions 17, 29, 40, 41, 42, 45, 47
energy efficiency 30, 31, 32, 33, 37, 42
Europe 8, 10, 14, 15, 16, 19, 20, 21, 22, 27, 28, 29, 33, 36, 37, 44

fossil fuels 5, 10, 11, 18, 21, 22, 23, 29, 39, 40, 43, 45, 47
fuel cells 35, 44, 45, 47

geothermal energy 16, 18, 44, 45, 47

global warming 5, 24, 25, 26, 27, 28, 29, 40, 41, 43, 47
greenhouse gases 25, 26, 28, 40, 42, 47

hybrid engines 32, 35, 42
hydroelectric power 13, 16, 18, 38, 44
hydrogen 35, 41, 45, 47

industry 6, 20, 21, 30, 31, 32, 40
insulation 31, 47

Kyoto agreement 40, 41, 42, 45

natural gas 11, 18, 21, 22, 23, 31, 44
nuclear fission 35, 47
nuclear fusion 35, 47
nuclear power 10, 11, 12, 18, 35, 40, 41, 45, 47

oil 8, 9, 11, 18, 19, 20, 21, 22, 23, 31, 33, 41, 44, 45

pollution 8, 9, 10, 11, 12, 25, 29, 32, 35

power plants 4, 8, 10, 13, 15, 16, 21, 30, 36
PV cells 15, 36, 37, 45, 47

renewable energy 10, 14, 15, 16, 17, 18, 34, 35, 42, 43, 47

solar cells (see PV cells)
solar power 10, 15, 31, 36, 37, 42, 44, 47

tidal power 15, 36
transport 6, 20, 25, 32, 33
turbines 13, 14, 15, 30, 34, 35, 36, 37, 38, 44, 47

UK 8, 15, 28, 33
United States 4, 5, 7, 8, 14, 15, 16, 19, 20, 21, 25, 26, 28, 33, 37, 40, 41, 44, 45

wave power 10, 15, 36
wildlife 14, 15, 28, 36, 38
wind power 10, 14, 36, 37, 38, 41, 42, 44
wood 6, 8, 11, 17, 21

Photo Credits:
Abbreviations: l-left, r-right, b-bottom, t-top, c-center, m-middle. Front cover, 5, 8, 11, 17t, 19, 25, 26tl, 29b—Photodisc. 1m, 42bl, 42br—BioRegional. 1l, 2tl, 18, 23tr—Comstock. 1r, 9, 23bl, 31br—Flat Earth. 2bl, 20, 32l, 40t, 44t—© European Community, 2005. 2-3, 38 (both), 45tr—© Elsam A/S. 3tr, 33t—Arup. 4—Associated Press. 6t, 10t, 13, 33b—Select Pictures. 6b, 26br, 31tl, 41b, 43b—Corbis. 7, 42t—Corel. 10bl—www.greenhouse.gov.au. 12—British Nuclear Fuels plc (BNFL). 14t—Aaron B Brown. 14b, 22, 27br, 39t, 40b, 44bl—Digital Vision. 15t—www.bp.com. 15b, 21, 35b—PBD. 16t—www.calpine.com. 16b—http://bioenergy.ornl.gov. 24 (all), 27t—NASA's Earth Observatory. 28b—Nicolas Benazeth. 30t—US Department of Energy. 32r—courtesy of The Lind Group. 34tr—EnviroMission. 35tr—www.mercedes-benz.com. 36t—Marine Current Turbines Ltd. 37t—Zapotec Energy, Cambridge, Mass., USA. 37br—BDSP Partnership. 43tl—European Parliament.

SALT SPRING ISLAND
MIDDLE SCHOOL